# What is a Shatter Cone?

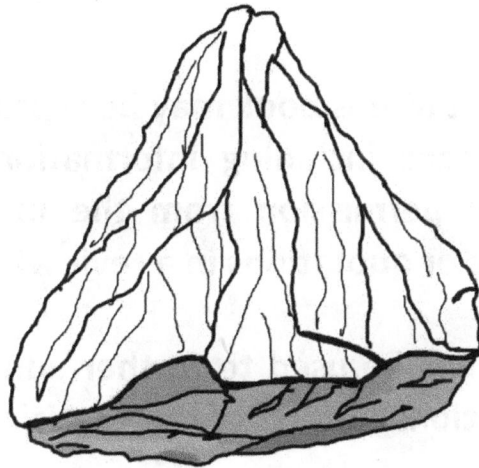

A Coloring Book by
The Georgia Mineral Society, Inc.

Written by Lori Carter

This edition published by:

The Georgia Mineral Society, Inc.
4138 Steve Reynolds Boulevard
Norcross, GA  30093-3059
www.gamineral.org

ISBN: 978-1-937617-03-5

# What is a shatter cone?

# Is
# it
# an
# ice-cream
# cone...

# that somebody dropped ?

# Is
# it
# a
# traffic
# cone...

The Georgia Mineral Society, Inc.

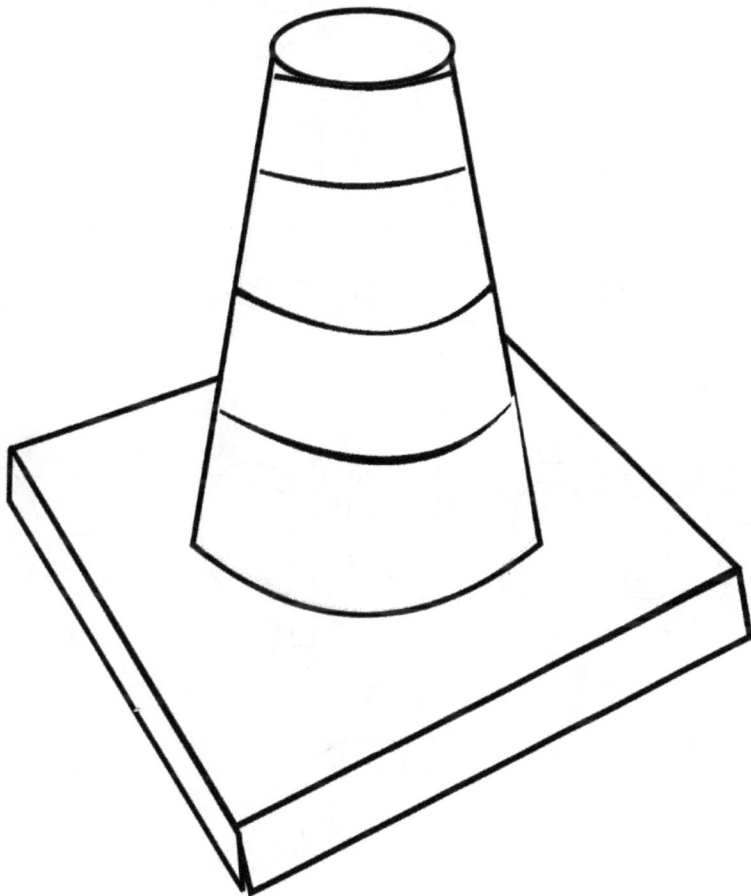

# where somebody stopped ?

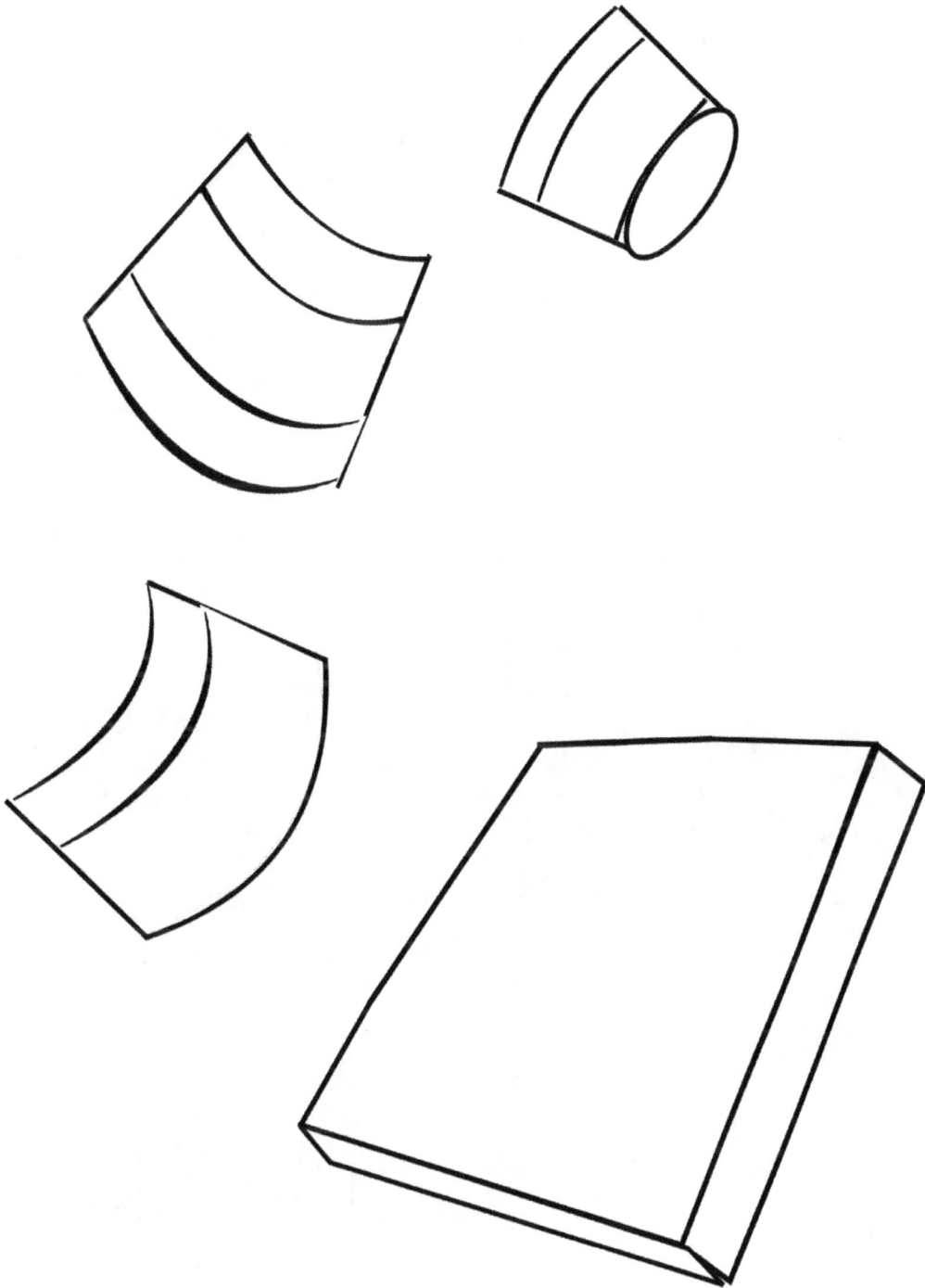

# No, it is a ROCK!

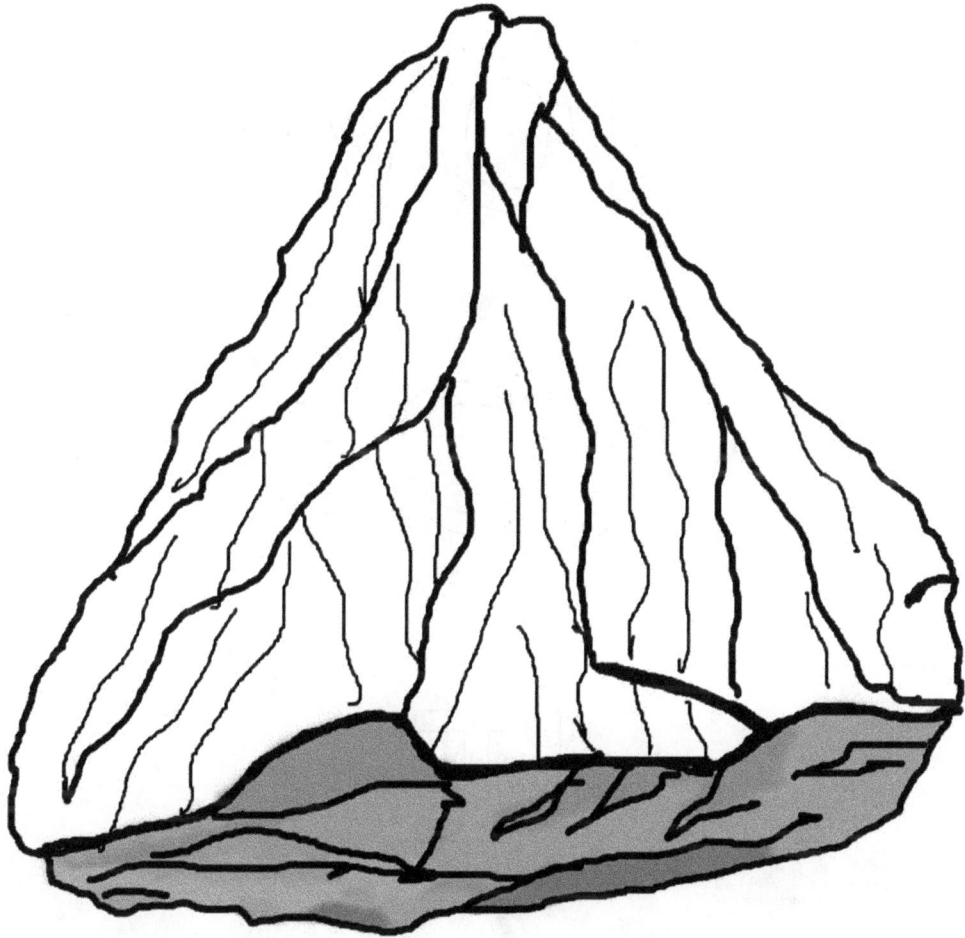

# A shatter cone is a rock!!!

# How was it made?

# Did it come from a volcano in a great big blast?

# Did it come from a dinosaur in a time long past?

# No. A thing from the sky hit the ground hard and fast!

POW!

# Did a big, heavy rain make the shatter cones?

# Did a terrible snow make the shatter cones?

# A meteorite from outer space
# Hit the earth and made a hole!

# It hit so hard it cracked some stones And made what we call shatter cones!

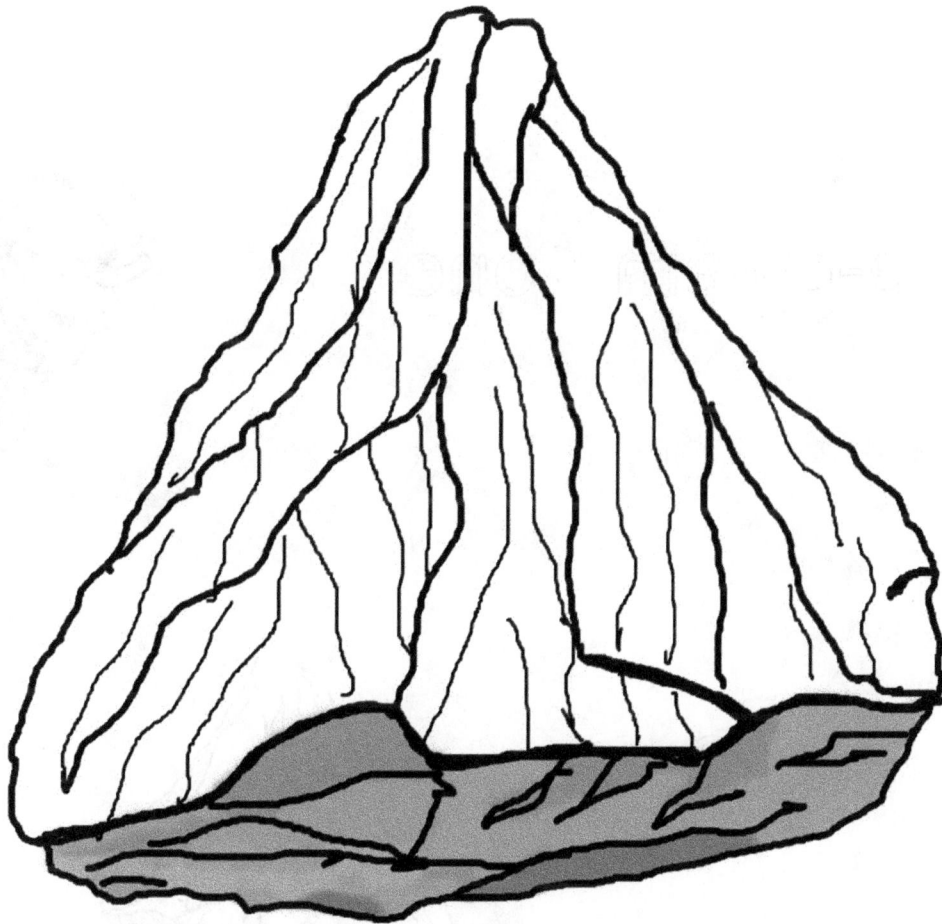

# Is a shatter cone

A Traffic Cone

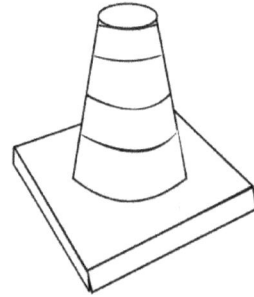

or

An Ice-cream Cone

or

A Rock?

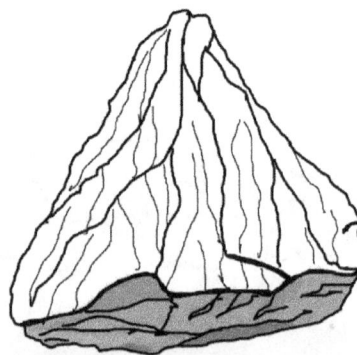

# Do shatter cones come from

Volcanoes

or

Dinosaurs

or

Something that came from space?

# What made the shatter cones?

Snow

or

A Meteorite

or

Rain?

# The
# End